WI

HAIRY FROG

AMY CULLIFORD

A Crabtree Roots Book

Crabtree Publishing
crabtreebooks.com

School-to-Home Support for Caregivers and Teachers

This book helps children grow by letting them practice reading. Here are a few guiding questions to help the reader with building his or her comprehension skills. Possible answers appear here in red.

Before Reading:

• What do I think this book is about?
  • *I think this book is about a weird frog with hair.*
  • *I think this book is about why the frog has hair.*

• What do I want to learn about this topic?
  • *I want to learn if the hairy frog is related to tree frogs.*
  • *I want to learn if the hairy frog can swim like some other frogs.*

During Reading:

• I wonder why...
  • *I wonder why some hairy frogs have body parts that look like hair.*
  • *I wonder why some hairy frogs lay eggs in water.*

• What have I learned so far?
  • *I have learned that hairy frogs have claws.*
  • *I have learned that hairy frogs can jump like other frogs.*

After Reading:

• What details did I learn about this topic?
  • *I have learned that hairy frogs live in big forests.*
  • *I have learned that hairy frogs lay their eggs in water.*

• Read the book again and look for the vocabulary words.
  • *I see the word **forests** on page 4 and the word **claws** on page 8. The other vocabulary words are found on page 14.*

Ribbit! I see a
hairy frog.

Hairy frogs live in big **forests**.

Some hairy frogs
have body parts
that look like **hair**.

All hairy frogs have **claws**.

Some hairy frogs lay **eggs** in water.

All hairy frogs
can jump!

Word List
Sight Words

a	in	see
all	jump	some
big	lay	that
can	like	water
have	live	
I	look	

Words to Know

claws

eggs

forests

hair

hairy frog

39 Words

Ribbit! I see a **hairy frog**.

Hairy frogs live in big **forests**.

Some hairy frogs have body parts that look like **hair**.

All hairy frogs have **claws**.

Some hairy frogs lay **eggs** in water.

All hairy frogs can jump!

WEIRD ANIMALS

HAIRY FROG

Written by: Amy Culliford

Designed by: Rhea Wallace

Series Development: James Earley

Proofreader: Melissa Boyce

Educational Consultant: Marie Lemke M.Ed.

Photographs:
Shutterstock: Paul Starosta: cover, p. 1, 11, 13; Joel Sartore:
    p. 3, 8-9; B. Fornius: p. 5; B. Trapp: p. 6-7;

Crabtree Publishing

crabtreebooks.com   800-387-7650

Copyright © 2024 Crabtree Publishing

All rights reserved. No part of this publication may be reproduced, stored in a retrieval system or be transmitted in any form or by any means, electronic, mechanical, photocopying, recording, or otherwise, without the prior written permission of Crabtree Publishing. In Canada: We acknowledge the financial support of the Government of Canada through the Canada Book Fund for our publishing activities.

Printed in the U.S.A./072023/CG20230214

Published in Canada
Crabtree Publishing
616 Welland Ave.
St. Catharines, Ontario
L2M 5V6

Published in the United States
Crabtree Publishing
347 Fifth Ave
Suite 1402-145
New York, NY 10016

Library and Archives Canada Cataloguing in Publication
Available at Library and Archives Canada

Library of Congress Cataloging-in-Publication Data
Available at the Library of Congress

Hardcover: 978-1-0398-0980-2
Paperback: 978-1-0398-1033-4
Ebook (pdf): 978-1-0398-1139-3
Epub: 978-1-0398-1086-0